SURVEYING MATHEMATICS MADE SIMPLE

An original book by

Jim Crume P.L.S., M.S., CFedS

Co-Authors
Cindy Crume
Bridget Crume
Troy Ray R.L.S.
Mark Sandwick L.S.I.T.

PRINTED EDITION

*****

PUBLISHED BY:
Jim Crume P.L.S., M.S., CFedS

The Myth about Spiral Curve Offsets

Book 7 of this Math-Series

Copyright 2013 © by Jim Crume P.L.S., M.S., CFedS

All Rights Reserved

First publication: November, 2013

Printed by CreateSpace

Available on Kindle and other devices

Cover photo courtesy of Ron Barbala, PLS-CFedS Geomatics Consulting Group

TERMS AND CONDITIONS

The content of the pages of this book is for your general information and use only. It is subject to change without notice.

Neither we nor any third parties provide any warranty or guarantee as to the accuracy, timeliness, performance, completeness or suitability of the information and materials found or offered in this book for any particular purpose. You acknowledge that such information and materials may contain inaccuracies or errors and we expressly exclude liability for any such inaccuracies or errors to the fullest extent permitted by law.

Your use of any information or materials in this book is entirely at your own risk, for which we shall not be liable. It shall be your own responsibility to ensure that any products, services or information available in this book meet your specific requirements.

This book may not be further reproduced or circulated in any form, including email. Any reproduction or editing by any means mechanical or electronic without the explicit written permission of Jim Crume is expressly prohibited.

Table of Contents

INTRODUCTION..4

THE MYTH ABOUT SPIRAL CURVE OFFSETS......7

SPIRAL OFFSET CURVE....................................12

POINTS ON A SPIRAL CURVE OFFSET................21

— METHOD ONE —..22

— METHOD TWO —..29

SOLUTIONS TO EXAMPLES................................42

CONCLUSION...53

ABOUT THE AUTHOR..54

INTRODUCTION

Straight forward Step-by-Step instructions.

This book is just one part in a series of digital and printed editions on Surveying Mathematics Made Simple. The subject matter in this book will utilize the methods and formulas that are covered in the books that precede it. If you have not read the preceding books, you are encouraged to review a copy before proceeding forward with this book.

For a list of books in this series, please visit:

http://www.cc4w.net/ebooks.html

Prerequisites for this book: A basic knowledge of geometry, algebra and trigonometry is required for the explanations shown in this book.

Book 1 - **Bearings and Azimuths** - How to add bearings and angles, subtract between bearings, convert from degrees-minutes-seconds to decimal degrees, convert from decimal degrees to degrees-minutes-seconds, convert from bearings to azimuths and convert from azimuths to bearings.

Book 2 - **Create Rectangular Coordinates** - How to calculate the northing and easting of an end point given the coordinates of the beginning point utilizing a bearing and distance of a line.

Book 3 - **Inverse Between Rectangular Coordinates** - How to determine the bearing and distance of a line given the coordinates for the beginning and ending point.

Book 4 - **Circular Curves** - How to calculate a circular curve, reverse curve, compound curve, Tangent In, Tangent Out and Local Tangent Bearing given only two parameters.

Book 5 - **Parcel Boundary** - Utilizing the lesson's learned in Book 1 through 4, Book 5 brings it all together to perform calculations for a parcel boundary and it's area.

Book 6 - **Spiral Curves** - How to calculate a centerline spiral curve with minimal known information, Equal Spiral Curves, Un-Equal Spiral Curves, the mythical Ten Chord Spiral and Points on a Spiral Curve.

Definitions:

Spiral Curve: (a.k.a. Euler Spiral, Easement Curve, Transition Curve) A curve which the degree of curvature at any point is proportional directly to the distance the point is from the point of zero curvature. The Euler Spiral curve is a clothoid with the cubic parabola having definite mathematical equations. This spiral curve is the accepted standard and has been utilized for highway and railroad curves since its inception by Leonard Euler.

Spiral Curve Offset: (a.k.a. Spiral Offset Curve) An offset curve that is parallel to a centerline spiral curve that is at an equal distance at any point along the offset curve measured perpendicular either left or right of the centerline spiral curve. The rate of change of the offset curve is directly related to the offset distance from the controlling centerline spiral curve.

THE MYTH ABOUT SPIRAL CURVE OFFSETS

One of my first projects straight out of technical school was laying out spiral curves on a railroad project from Bill, Wyoming to Douglas, Wyoming using a Wild T2 theodolite and a 100' metal chain. Not having been formally trained on calculating spiral curves or calculus for that matter, I purchased a book titled "Railroad Curves and Earthwork" by C. Frank Allen, S.B. that gave me the necessary formulas needed to determine spiral curves for this railroad project. This book was expensive and hard to find back in those days, but worth every penny.

This book instructed me on the Ten Chord Spiral curve concept that the railroad uses and the process to properly layout spiral curves once the Tangent In and Tangent Out have been determined and staked in the field.

Staking the centerline was not a major challenge. The challenge came in staking the right-of-way line left and right parallel to the centerline spiral curve for the fencing crew.

This book did not have any formulas for calculating a true parallel offset curve to a centerline spiral curve.

The method described in the book was to:

1) Stake the spiral centerline in ten equal chords, then

2) Occupy the points along the centerline spiral curve, then

3) Turn the appropriate angle to get perpendicular to the centerline, and

4) Measure out the offset distance left and right to the right of way line.

This method was slow and time-consuming. Where there was a significant vertical difference from the centerline to the right-of-way line, slope distances were required.

Back then, there were no methods or formulas for staking the spiral offset curve from the right-of-way tangent line. In 1977 the survey equipment consisted of theodolites, steel chains, and a large (big orange box) electronic distance measurement device (EDM) that was awkward to use. The company I worked for only had one EDM. It was used for the initial vertical and horizontal control but was not available for the construction staking we were doing. The only equipment available at the time was a Wild T2 and T16 theodolite, a 100-foot and a 300-foot steel chain. Using today's survey equipment (GPS/total station) would have made the work so much easier.

Because I like math, I spent many hours at night in the hotel room calculating by hand coordinates for the spiral offset points that could be staked radially from a nearby control point (personal computers had not been invented yet). I did this while the crew spent their time at the local bar. Back in those days performing a radial stakeout was frowned upon by the surveying profession. I tend to venture outside of the comfort zone of most survey professionals. Ironically, today radial stakeout is the most widely used method.

The longest distance that could be measured was 300 feet from a control point. Therefore, many intermediate control points were set from the major control point network that was established previously with the EDM and theodolite.

I became friends with a professional engineer on the project that worked for Burlington Northern Railroad, and together we spent many hours trying to come up with a process of calculating and staking the offset spiral curve using the same method that was used for the centerline. We both struggled to come up with the formulas with no success.

Over the years I have researched many surveying books, talked to many surveyors and engineers and the common belief was

"There is no such a thing as a true parallel offset curve to a centerline spiral curve".

I went along with that common belief for many years. Deep inside me I felt strongly that there had to be a solution.

This Myth has finally been busted.

After working on a solution starting in 1977, I have finally derived a method and formulas that will calculate a spiral offset curve that is truly parallel to the centerline spiral.

The centerline spiral curves in this book are based upon the Euler (Pronounced 'oil-er') Spiral Curve developed by Leonhard Euler. The Euler Spiral Curve is also the same mathematical model that is in use on most highway curves today.

Keep in mind that there maybe transportation agencies that may use other mathematical models of spiral curves that are not covered in this book. I personally have not ran across any use of spiral curves other than the Euler Spiral mathematical model.

NOTES

SPIRAL OFFSET CURVE

Figure 1 shows the various components for a spiral curve offset. It is important that you become familiar with these components. They will be referenced throughout this book.

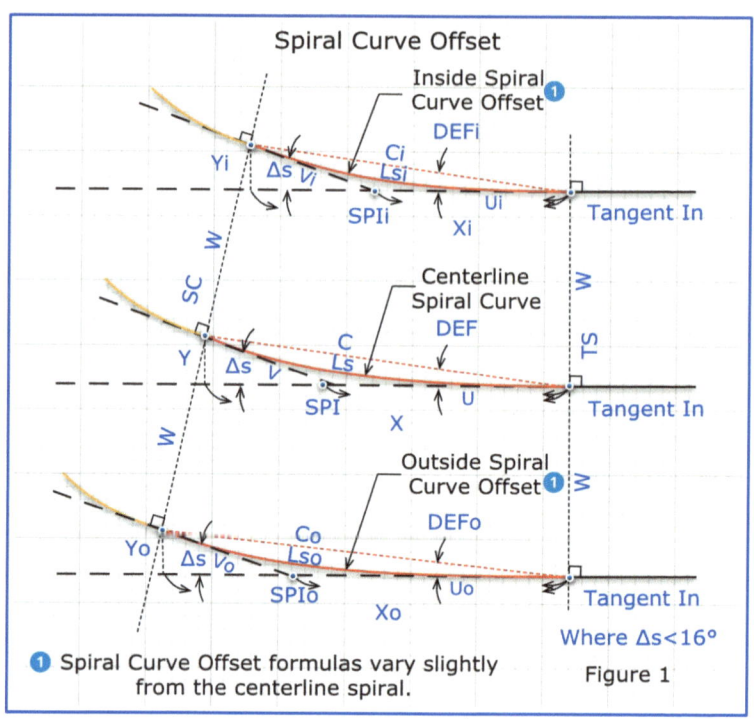

Figure 1

Definitions:

W = Spiral Offset distance

Xi, Xo = Distance along the Spiral Curve Offset Tangent In

Yi, Yo = Perpendicular offset from the Spiral Curve Offset Tangent In

Ci, Co = Spiral Offset Chord

Vi, Vo = Distance along Local Tangent to the SC(Sta)

Ui, Uo = Distance along the Tangent In to the Spiral Offset PI

DEFi, DEFo = Spiral Offset Deflection angle

Lsi, Lso = Spiral Offset Length

SPIi, SPIo = Spiral Offset PI

Ri, Ro = Radius of Main Curve at Spiral Offset distance

Di, Do = Degree of Curvature of Main Curve at Spiral Offset distance

ai, ao = Rate of change per 100' along Spiral Offset

Spiral Offset Curve characteristics vary slightly from the centerline spiral curve. The Spiral Curve Offset formulas presented herein are dependent upon the centerline spiral curve. You must have a thorough understanding of the centerline spiral curve before proceeding with this book.

See Book 6 "**Spiral Curves**" for definitions and formulas for the centerline spiral curve.

Formulas:

—Inside Spiral Curve Offset—-

$Xi = X - (Sin(\Delta s) * W)$

$Yi = Y - W + (Cos(\Delta s) * W)$

$Ci = \sqrt{(Xi^2 + Yi^2)}$

$Vi = Yi / Sin(\Delta s)$

$U_i = X_i - (Y_i / \operatorname{Tan}(\Delta s))$

$L_{si} = C_i * L_s / C$

$DEF_i = \operatorname{ArcTan}(Y_i / X_i)$

$R_i = R - W$

$D_i = 5729.57795 / R_i$ (Arc definition)

$a_i = D_i * 100 / L_{si}$

—Outside Spiral Curve Offset—-

$X_o = X + (\operatorname{Sin}(\Delta s) * W)$

$Y_o = Y + W - (\operatorname{Cos}(\Delta s) * W)$

$C_o = \sqrt{(X_o^2 + Y_o^2)}$

$V_o = Y_o / \operatorname{Sin}(\Delta s)$

$U_o = X_o - (Y_o / \operatorname{Tan}(\Delta s))$

$L_{so} = C_o * L_s / C$

$DEF_o = \operatorname{ArcTan}(Y_o / X_o)$

$R_o = R + W$

$D_o = 5729.57795 / R_o$ (Arc definition)

$a_o = D_o * 100 / L_{so}$

Note: Rounding error is dependent upon the number of decimal places that are utilized. It is recommended that at least 5 decimal places be used for all calculations then round the final answer as needed.

All angles must be converted to Decimal Degrees prior to performing trigonometric operations. See Book 1 - "Bearings and Azimuths" for methods on converting Degrees-Minutes-Seconds to Decimal Degrees and vice versa. Also see Book 1 for adding and subtracting bearings and angles.

See Book 4 - "Circular Curves" for definitions for "Arc definition" and "Chord definition".

Example 1:

Given:

—Centerline Spiral Curve—

$\Delta s = 2°00'00"$

$Ls = 200.00000$

$C = 199.98912$

$R = 2864.78898$

$X = 199.97558$

$Y = 2.32693$

$W = 100.00$

See Book 6 - "Spiral Curves" for definitions for the centerline spiral curve.

Solve for the following elements:

—Inside Spiral Curve Offset—

$Xi = X - (\mathrm{Sin}(\Delta s) * W)$

$Xi = 199.97558 - (\mathrm{Sin}(2°00'00") * 100.00)$

$Xi = \mathbf{196.48563}$

$Yi = Y - W + (\mathrm{Cos}(\Delta s) * W)$

$Yi = 2.32693 - 100.00 + (\mathrm{Cos}(2°00'00") * 100.00)$

$Yi = \mathbf{2.26601}$

$Ci = \sqrt{(Xi^2 + Yi^2)}$

$Ci = \sqrt{(196.48563^2 + 2.26601^2)}$

$Ci = \mathbf{196.49870}$

$Vi = Yi / Sin(\Delta s)$

$Vi = 2.26601 / Sin(2°00'00")$

$Vi = \mathbf{64.92959}$

$Ui = Xi - (Yi / Tan(\Delta s))$

$Ui = 196.48563 - (2.26601 / Tan(2°00'00"))$

$Ui = \mathbf{131.59559}$

$Lsi = Ci * Ls / C$

$Lsi = 196.49870 * 200.00000 / 199.98912$

$Lsi = \mathbf{196.50939}$

$DEFi = ArcTan(Yi / Xi)$

$DEFi = ArcTan(2.26601 / 196.48563)$

$DEFi = \mathbf{0.66075° \text{ or } 00°39'39"}$

$Ri = R - W$

$Ri = 2864.78898 - 100.00$

$Ri = \mathbf{2764.78898}$

$Di = 5729.57795 / Ri$

$Di = 5729.57795 / 2764.78898$

$Di = \mathbf{2.07234° \text{ or } 2°04'20"}$

$ai = Di * 100 / Lsi$

$ai = 2.07234 * 100.00 / 196.50939$

$ai = \mathbf{1.05458}$

—Outside Spiral Curve Offset—-

$Xo = X + (Sin(\Delta s) * W)$

$Xo = 199.97558 + (Sin(2°00'00") * 100.00)$

$X_o = \mathbf{203.46553}$

$Y_o = Y + W - (\cos(\Delta s) * W)$

$Y_o = 2.32693 + 100.00 - (\cos(2°00'00'') * 100.00)$

$Y_o = \mathbf{2.38785}$

$C_o = \sqrt{(X_o^2 + Y_o^2)}$

$C_o = \sqrt{(203.46553^2 + 2.38785^2)}$

$C_o = \mathbf{203.47954}$

$V_o = Y_o / \sin(\Delta s)$

$V_o = 2.38785 / \sin(2°00'00'')$

$V_o = \mathbf{68.42076}$

$U_o = X_o - (Y_o / \tan(\Delta s))$

$U_o = 203.46553 - (2.38785 / \tan(2°00'00''))$

$U_o = \mathbf{135.08645}$

$L_{so} = C_o * L_s / C$

$L_{so} = 203.47954 * 200.00000 / 199.98912$

$L_{so} = \mathbf{203.49061}$

$DEF_o = \arctan(Y_o / X_o)$

$DEF_o = \arctan(2.38785 / 203.46553)$

$DEF_o = \mathbf{0.67239° \text{ or } 00°40'21''}$

$R_o = R + W$

$R_o = 2864.78898 + 100.00$

$R_o = \mathbf{2964.78898}$

$D_o = 5729.57795 / R_o$

$D_o = 5729.57795 / 2964.78898$

$D_o = \mathbf{1.93254° \text{ or } 1°55'57''}$

ao = Do * 100 / Lso

ao = 1.93254 * 100 / 203.49061

ao = **0.94969**

The following table lists values for the various components for the centerline spiral curve and spiral offset curves, inside and outside at offset distances of 50' and 100'.

	Spiral Curve Offset Table						
	W	a	D	R	Length	Delta	DEF
Inside	100	1.05458	2°04'20"	2764.78898	196.50939	2°00'00"	00°39'39"
	50	1.02678	2°02'08"	2814.78898	198.24391	2°00'00"	00°39'49"
Centerline	0	1	2°00'00"	2864.78898	200.00000	2°00'00"	00°40'00"
Outside	50	0.97440	1°57'57"	2914.78898	201.73433	2°00'00"	00°40'10"
	100	0.94969	1°55'57"	2964.78898	203.49061	2°00'00"	00°40'21"
	W	Chord	X	Y	U	V	
Inside	100	196.49870	196.48563	2.26601	131.59559	64.92959	
	50	198.24391	198.23061	2.29647	132.46837	65.80232	
Centerline	0	199.98912	199.97558	2.32693	133.34112	66.67508	
Outside	50	201.73433	201.72056	2.35739	134.21388	67.54783	
	100	203.47954	203.46553	2.38785	135.08645	68.42076	

NOTES

Practical Example 1

The south line of a private parcel is in common with the existing north right of way line of a state highway. The legal description describes this common line as being parallel and 75.00 feet right, measured perpendicular, to the centerline spiral curve. The deflection angle, spiral offset curve length and chord length is required for the common line. The common line is an Inside Spiral Curve Offset.

Given:

—Centerline Spiral Curve—-

$\Delta s = 2°00'00"$

$Ls = 200.00000$

$C = 199.98912$

$R = 2864.78898$

$X = 199.97558$

$Y = 2.32693$

$W = 75.00$

Solve for the following elements:

—Inside Spiral Curve Offset—-

$DEFi = ??°??'??"$

$Ci = ???.?????$

$Lsi = ???.?????$

The solution can be found at the end of the book.

POINTS ON A SPIRAL CURVE OFFSET

Calculating points on a spiral curve offset can be done using one of three methods.

Method One is by determining the distance along the tangent line of the centerline spiral curve then determining the perpendicular offset distance to the Point on the Spiral Curve Offset.

Method Two is by determining the distance along the tangent line of the spiral curve offset then determining the perpendicular offset distance to the Point on the Spiral Curve Offset.

Method Three is performed by determining a line that is perpendicular to the Point on the centerline Spiral Curve then calculating the offset distance along the perpendicular line. This method is depicted in most technical manuals on this subject.

Methods One and Two are new methods that I have developed after many years of trying to find a better way of calculating points on a spiral curve offset.

Methods One and Two will be described in this book.

— METHOD ONE —

Figure 2 shows the various components for points on a spiral curve offset. It is important that you become familiar with these components. They will be referenced throughout this book.

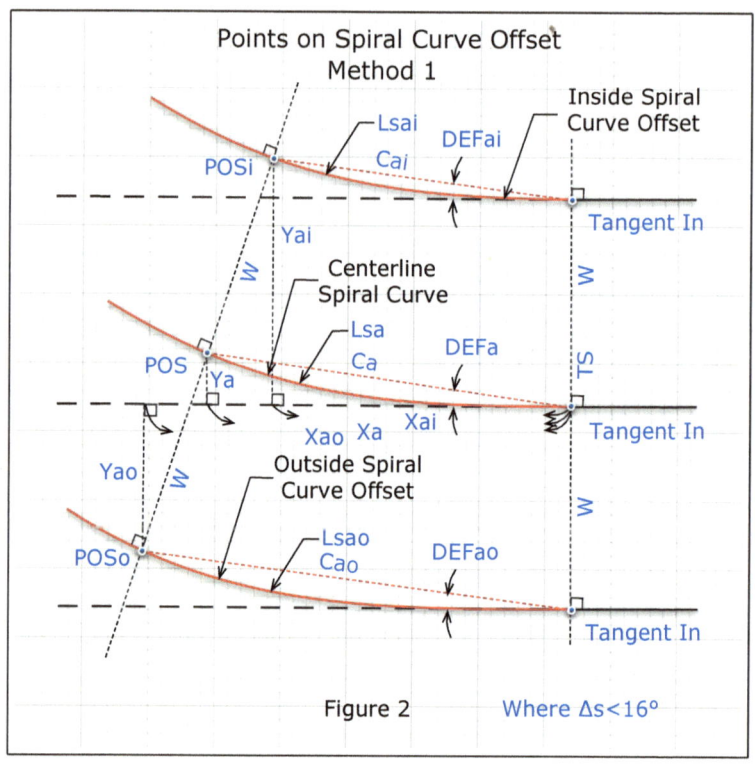

Figure 2

Definitions:

W = Spiral Offset distance

Xa, Xai, Xao = Distance along the Spiral Curve Tangent In

Ya, Yai, Yao = Perpendicular offset from the Spiral Curve Tangent In

Ca, Cai, Cao = Spiral & Spiral Offset Chord

DEFa, DEFai, DEFao = Spiral & Spiral Offset Deflection angle

Lsa, Lsai, Lsao = Spiral & Spiral Offset Length

a = Rate of change per 100' along the centerline Spiral Curve

Formulas:

—Centerline Spiral Curve—-

Lsa = POS(sta) - TS(Sta)

Ca = Lsa - (0.00034 * a² * (Lsa / 100)⁵)

DEFa = (a *Lsa²) / 60000

Xa = Ca * Cos(DEFa)

Ya = Ca * Sin(DEFa)

—Inside Spiral Curve Offset—-

Xai = Xa - (Sin(DEFa * 3) * W)

Yai = Ya + (Cos(DEFa * 3) * W)

Cai = √(Xai² + (Yai - W)²)

Lsai = Cai * Lsa / Ca

DEFai = ArcTan((Yai - W) / Xai)

—Outside Spiral Curve Offset—-

Xao = Xa + (Sin(DEFa * 3) * W)

Yao = Ya - (Cos(DEFa * 3) * W)

Cao = √(Xao² + (Yao + W)²)

Lsao = Cao * Lsa / Ca

$DEF_{ao} = ArcTan((Y_{ao} + W) / X_{ao})$

Note: Rounding error is dependent upon the number of decimal places that are utilized. It is recommended that at least 5 decimal places be used for all calculations then round the final answer as needed.

All angles must be converted to Decimal Degrees prior to performing trigonometric operations. See Book 1 - "Bearings and Azimuths" for methods on converting Degrees-Minutes-Seconds to Decimal Degrees and vice versa. Also see Book 1 for adding and subtracting bearings and angles.

Example 2:

Given:

—**Centerline Spiral Curve**—

a = 1.00000

TS(Sta) = 2180+84.70

POS(Sta) = 2182+00.00

W = 100.00

Solve for the following elements:

—**Centerline Spiral Curve**—

Lsa = POS(sta) - TS(Sta)

Lsa = 2182+00.00 - 2180+84.70

Lsa = **115.30**

Ca = Lsa - (0.00034 * a^2 * (Lsa / 100)^5)

Ca = 115.30 - (0.00034 * 1.00000^2 * (115.30 / 100)^5)

Ca = **115.29931**

DEFa = (a *Lsa^2) / 60000

DEFa = (1.00000 * 115.30^2) / 60000

DEFa = **0.22157° or 0°13'18"**

Xa = Ca * Cos(DEFa)

Xa = 115.29931 * Cos(0°13'18")

Xa = **115.29845**

Ya = Ca * Sin(DEFa)

Ya = 115.29931 * Sin(0°13'18")

Ya = **0.44588**

—Inside Spiral Curve Offset—-

Xai = Xa - (Sin(DEFa * 3) * W)

Xai = 115.29845 - (Sin(0°13'18" * 3) * 100.00)

Xai = **114.13834**

Yai = Ya + (Cos(DEFa * 3) * W)

Yai = 0.44588 + (Cos(0°13'18" * 3) * 100.00)

Yai = **100.43915**

Cai = √(Xai² + (Yai - W)²)

Cai = √(114.13834² + (100.43915 - 100.00)²)

Cai = **114.13918**

Lsai = Cai * Lsa / Ca

Lsai = 114.13918 * 115.30 / 115.29931

Lsai = **114.13986**

DEFai = ArcTan((Yai - W) / Xai)

DEFai = ArcTan((100.43915 - 100.00) / 114.13834

DEFai = **0.22045° or 0°13'14"**

—Outside Spiral Curve Offset—-

Xao = Xa + (Sin(DEFa * 3) * W)

Xao = 115.29845 + (Sin(0°13'18" * 3) * 100.00)

Xao = **116.45856**

Yao = Ya - (Cos(DEFa * 3) * W)

Yao = 0.44588 - (Cos(0°13'18" * 3) * 100.00)

Yao = **-99.54739**

$Cao = \sqrt{(Xao^2 + (Yao + W)^2)}$

$Cao = \sqrt{(116.45856^2 + (-99.54739 + 100.00)^2)}$

Cao = **116.45944**

Lsao = Cao * Lsa / Ca

Lsao = 116.45944 * 115.30 / 115.29931

Lsao = **116.46014**

DEFao = ArcTan((Yao + W) / Xao

DEFao = ArcTan((-99.54739 + 100.00) / 116.45856

DEFao = **0.22268° or 0°13'22"**

NOTES

— METHOD TWO —

Figure 3 shows the various components for points on a spiral curve offset. It is important that you become familiar with these components. They will be referenced throughout this book.

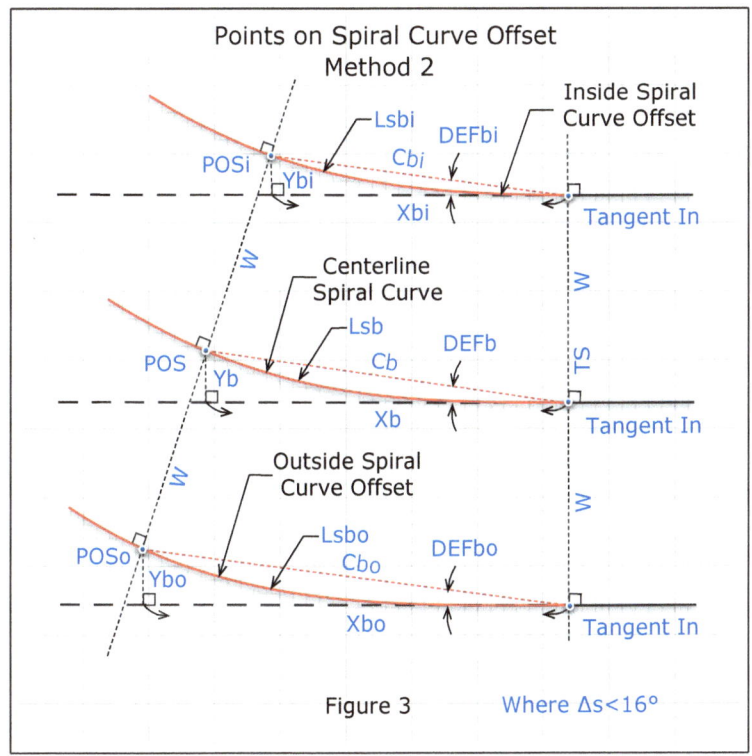

Figure 3

Definitions:

W = Spiral Offset distance

Xb, Xbi, Xbo = Distance along the Spiral & Spiral Offset Tangent In

Yb, Ybi, Ybo = Perpendicular offset from the Spiral & Spiral Offset Tangent In

Cb, Cbi, Cbo = Spiral & Spiral Offset Chord

DEFb, DEFbi, DEFbo = Spiral & Spiral Offset Deflection angle

Lsb, Lsbi, Lsbo = Spiral & Spiral Offset Length

a = Rate of change per 100' along the centerline Spiral Curve

Formulas:

—Centerline Spiral Curve—-

Lsb = POS(sta) - TS(Sta)

Cb = Lsb - (0.00034 * a^2 * (Lsb / 100)^5)

DEFb = (a *Lsb^2) / 60000

Xb = Cb * Cos(DEFb)

Yb = Cb * Sin(DEFb)

—Inside Spiral Curve Offset—-

Xbi = Xb - (Sin(DEFb * 3) * W)

Ybi = Yb + (Cos(DEFb * 3) * W) - W

Cbi = √(Xbi^2 + Ybi^2)

Lsbi = Cbi * Lsb / Cb

DEFbi = ArcTan(Ybi / Xbi)

—Outside Spiral Curve Offset—-

Xbo = Xb + (Sin(DEFb * 3) * W)

Ybo = Yb + W - (Cos(DEFb * 3) * W)

Cbo = √(Xbo^2 + Ybo^2)

Lsbo = Cbo * Lsb / Cb

DEFbo = ArcTan(Ybo / Xbo)

Note: Rounding error is dependent upon the number of decimal places that are utilized. It is recommended that at least 5 decimal places be used for all calculations then round the final answer as needed.

All angles must be converted to Decimal Degrees prior to performing trigonometric operations. See Book 1 - "Bearings and Azimuths" for methods on converting Degrees-Minutes-Seconds to Decimal Degrees and vice versa. Also see Book 1 for adding and subtracting bearings and angles.

NOTES

Example 3:

Given:

—**Centerline Spiral Curve**—

a = 1.00000

TS(Sta) = 2180+84.70

POS(Sta) = 2182+50.00

W = 100.00

Solve for the following elements:

—**Centerline Spiral Curve**—

Lsb = POS(sta) - TS(Sta)

Lsb = 2182+50.00 - 2180+84.70

Lsb = **165.30**

Cb = Lsb - (0.00034 * a^2 * (Lsb / 100)^5)

Cb = 165.30 - (0.00034 * 1.00000^2 * (165.30 / 100)^5)

Cb = **165.29580**

DEFb = (a *Lsb^2) / 60000

DEFb = (1.00000 * 165.30^2) / 60000

DEFb = **0.45540° or 0°27'19"**

Xb = Cb * Cos(DEFb)

Xb = 165.29580 * Cos(0°27'19")

Xb = **165.29058**

Yb = Cb * Sin(DEFb)

Yb = 165.29580 * Sin(0°27'19")

Y_b = **1.31380**

—**Inside Spiral Curve Offset**—

X_{bi} = X_b - (Sin(DEF$_b$ * 3) * W)

X_{bi} = 165.29058 - (Sin(0°27'19" * 3) * 100.00)

X_{bi} = **162.90634**

Y_{bi} = Y_b + (Cos(DEF$_b$ * 3) * W) - W

Y_{bi} = 1.31380 + (Cos(0°27'19" * 3) * 100.00) - 100.00

Y_{bi} = **1.28537**

C_{bi} = √(X_{bi}^2 + Y_{bi}^2)

C_{bi} = √(162.90634² + 1.28537²)

C_{bi} = **162.91141**

L_{sbi} = C_{bi} * L_{sb} / C_b

L_{sbi} = 162.91141 * 165.30 / 165.29580

L_{sbi} = **162.91555**

DEF$_{bi}$ = ArcTan(Y_{bi} / X_{bi})

DEF$_{bi}$ = ArcTan(1.28537 / 162.90634)

DEF$_{bi}$ = **0.45207° or 0°27'07"**

—**Outside Spiral Curve Offset**—

X_{bo} = X_b + (Sin(DEF$_b$ * 3) * W)

X_{bo} = 165.29058 + (Sin(0°27'19" * 3) * 100.00)

X_{bo} = **167.67482**

Y_{bo} = Y_b + W - (Cos(DEF$_b$ * 3) * W)

Y_{bo} = 1.31380 + 100.00 - (Cos(0°27'19" * 3) * 100.00)

$Y_{bo} = \mathbf{1.34223}$

$C_{bo} = \sqrt{(X_{bo}^2 + Y_{bo}^2)}$

$C_{bo} = \sqrt{(167.67482^2 + 1.34223^2)}$

$C_{bo} = \mathbf{167.68019}$

$L_{sbo} = C_{bo} * L_{sb} / C_b$

$L_{sbo} = 167.67882 * 165.30 / 165.29580$

$L_{sbo} = \mathbf{167.68308}$

$DEF_{bo} = ArcTan(Y_{bo} / X_{bo})$

$DEF_{bo} = ArcTan(1.34223 / 167.67482)$

$DEF_{bo} = \mathbf{0.45864° \text{ or } 0°27'31''}$

NOTES

Practical Example 2

Your client has requested that you set property corners along the existing highway right of way line in common with their south property line. A segment of their property line parallel's the centerline spiral curve per the legal description. There is also a jog in this right of way line near the middle of the spiral offset curve. **Figure 4** shows the parallel spiral curve offset for the existing common line.

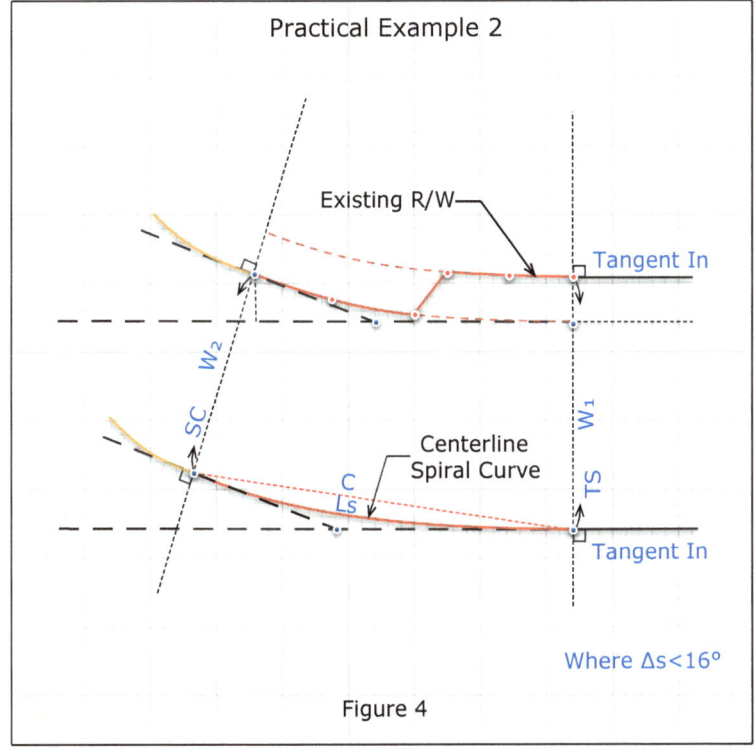

Figure 4

Figure 5 shows the points that the client has requested be set along the spiral offset curve. Points

1, 3, 4 and 6 will be permanent monuments. Points 2 & 5 will be temporary points.

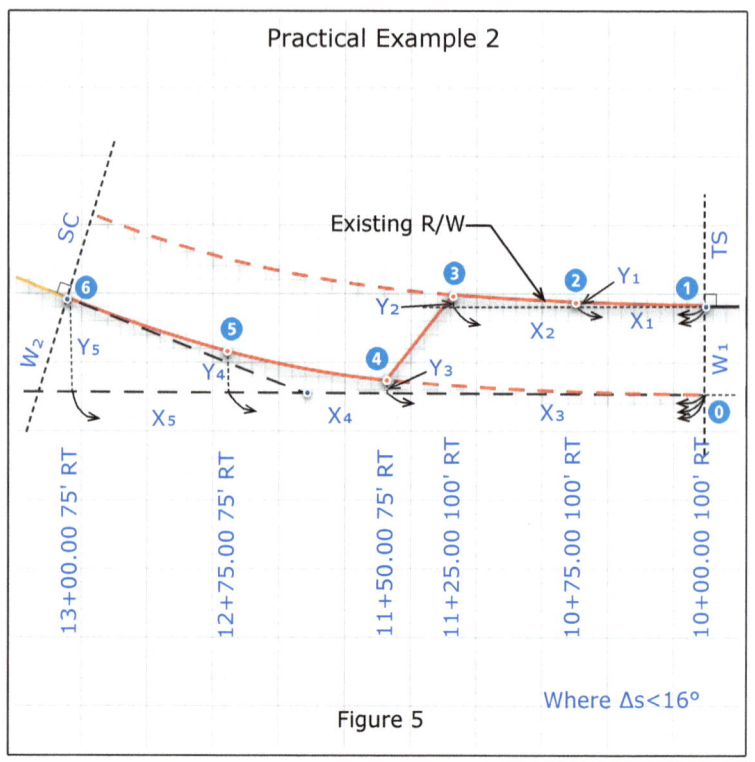

Figure 5

Using **Method 2**, that was previously described, calculate coordinates for the points along the spiral offset curve per the station and offsets shown in **Figure 5** that can be utilized to stake these points with a GPS.

Hint: Solve each spiral curve offset separately for W_1 and W_2.

Given:

—Centerline Spiral Curve—-

a = 0.66667

Ls = 300.00

TS(Sta) = 10+00.00

TS N = 10000.00000

TS E = 20000.00000

Tangent In = N90°00'00"W

—Offsets—-

W_1 = 100.00

W_2 = 75.00

Solve for the following elements:

—Inside Spiral Curve Offset—-

N_1 = ?????.?????

E_1 = ?????.?????

N_2 = ?????.?????

E_2 = ?????.?????

N_3 = ?????.?????

E_3 = ?????.?????

N_4 = ?????.?????

E_4 = ?????.?????

N_5 = ?????.?????

E_5 = ?????.?????

$N_6 = ?????.?????$

$E_6 = ?????.?????$

The solution can be found at the end of the book.

NOTES

SOLUTIONS TO EXAMPLES

Note: Rounding error is dependent upon the number of decimal places that are utilized. It is recommended that at least 5 decimal places be used for all calculations then round the final answer as needed.

All angles must be converted to Decimal Degrees prior to performing trigonometric operations. See Book 1 - "Bearings and Azimuths" for methods on converting Degrees-Minutes-Seconds to Decimal Degrees and vice versa. Also see Book 1 for adding and subtracting bearings and angles.

Solution for Practical Example 1:

Given:

—**Centerline Spiral Curve**—

$\Delta s = 2°00'00"$

$Ls = 200.00000$

$C = 199.98912$

$R = 2864.78898$

$X = 199.97558$

$Y = 2.32693$

$W = 75.00$

Solve for the following elements:

—**Inside Spiral Curve Offset**—

$DEFi = ??°??'??"$

$\quad Xi = X - (Sin(\Delta s) * W)$

$\quad Xi = 199.97558 - (Sin(2°00'00") * 75.00)$

$X_i = \mathbf{197.35812}$

$Y_i = Y - W + (\cos(\Delta s) * W)$

$Y_i = 2.32693 - 75.00 + (\cos(2°00'00'') * 75.00)$

$Y_i = \mathbf{2.28124}$

$DEF_i = \text{ArcTan}(Y_i / X_i)$

$DEF_i = \text{ArcTan}(2.28124 / 197.35812)$

$DEF_i = \mathbf{0.66225° \text{ or } 00°39'44''}$

$C_i = ???.?????$

$C_i = \sqrt{(X_i^2 + Y_i^2)}$

$C_i = \sqrt{(197.35812^2 + 2.28124^2)}$

$C_i = \mathbf{197.37130}$

$L_{si} = ???.?????$

$L_{si} = C_i * L_s / C$

$L_{si} = 197.37130 * 200.00000 / 199.98912$

$L_{si} = \mathbf{197.38204}$

Solution for Practical Example 2:

Given:

—Centerline Spiral Curve—-

a = 0.66667

LS = 300.00

TS(Sta) = 10+00.00

TS N = 10000.00000

TS E = 20000.00000

Tangent In = N90°00'00"W

—Offsets—-

W_1 = 100.00

W_2 = 75.00

Solve for the following elements:

—Inside Spiral Curve Offset at 100.00 RT—-
—Point 1—-

Note: The "Tangent In" is N90°00'00"W therefore to calculate the northing for Point 1 use simple addition.

N_1 = ?????.?????

    N_1 = TS N + W_1

    N_1 = 10000.00000 + 100.00

    N_1 = **10100.00000**

E_1 = ?????.?????

    E_1 = TS E

$E_1 = \mathbf{20000.00000}$

—Point 2—-

Lsb = POS(sta) - TS(Sta)

Lsb = 10+75.00 - 10+00.00

Lsb = **75.00**

Cb = Lsb - (0.00034 * a² * (Lsb / 100)⁵)

Cb = 75.00 - (0.00034 * 0.66667² * (75.00 / 100)⁵)

Cb = **74.99996**

DEFb = (a *Lsb²) / 60000

DEFb = (0.66667 * 75.00²) / 60000

DEFb = **0.06250° or 0°03'45"**

Xb = Cb * Cos(DEFb)

Xb = 74.99996 * Cos(0°03'45")

Xb = **75.99992**

Yb = Cb * Sin(DEFb)

Yb = 74.99996 * Sin(0°03'45")

Yb = **0.08181**

N_2 = ?????.?????

    Y_1 = Yb + (Cos(DEFb * 3) * W_1) - W_1

    Y_1 = 0.08181 + (Cos(0°03'45" * 3) * 100.00) - 100.00

    Y_1 = **0.08127**

    $N_2 = N_1 + Y_1$

    N_2 = 10100.00000 + 0.08127

$N_2 = \mathbf{10100.08127}$

$E_2 = ?????.?????$

    $X_1 = X_b - (\text{Sin}(DEF_b * 3) * W_1)$

    $X_1 = 74.99992 - (\text{Sin}(0°03'45" * 3) * 100.00)$

    $X_1 = \mathbf{74.67267}$

    $E_2 = E_1 - X_1$

    $E_2 = 20000.00000 - 74.67267$

    $E_2 = \mathbf{19925.32733}$

—**Point 3**—

$L_{sb} = \text{POS(sta)} - \text{TS(Sta)}$

$L_{sb} = 11+25.00 - 10+00.00$

$L_{sb} = \mathbf{125.00}$

$C_b = L_{sb} - (0.00034 * a^2 * (L_{sb} / 100)^5)$

$C_b = 125.00 - (0.00034 * 0.66667^2 * (125.00 / 100)^5)$

$C_b = \mathbf{124.99954}$

$DEF_b = (a * L_{sb}^2) / 60000$

$DEF_b = (0.66667 * 125.00^2) / 60000$

$DEF_b = \mathbf{0.17361° \text{ or } 0°10'25"}$

$X_b = C_b * \text{Cos}(DEF_b)$

$X_b = 124.99954 * \text{Cos}(0°10'25")$

$X_b = \mathbf{124.99897}$

$Y_b = C_b * \text{Sin}(DEF_b)$

$Y_b = 124.99954 * \text{Sin}(0°10'25")$

$Y_b = \mathbf{0.37876}$

N_3 = ?????.?????

$\quad Y_2 = Y_b + (\cos(DEF_b * 3) * W_1) - W_1$

$\quad Y_2 = 0.37876 + (\cos(0°10'25" * 3) * 100.00) - 100.00$

$\quad Y_2 = $ **0.37463**

$\quad N_3 = N_1 + Y_2$

$\quad N_3 = 10100.00000 + 0.37463$

$\quad N_3 = $ **10100.37463**

E_3 = ?????.?????

$\quad X_2 = X_b - (\sin(DEF_b * 3) * W_1)$

$\quad X_2 = 124.99897 - (\sin(0°10'25" * 3) * 100.00)$

$\quad X_2 = $ **124.08996**

$\quad E_3 = E_1 - X_2$

$\quad E_3 = 20000.00000 - 124.08996$

$\quad E_3 = $ **19875.91004**

—**Inside Spiral Curve Offset at 75.00 RT**—
—**Point 0**—

Note: The "Tangent In" is N90°00'00"W therefore to calculate the northing for Point 0 use simple addition.

N_0 = ?????.?????

$\quad N_0 = TS\ N + W_1$

$\quad N_0 = 10000.00000 + 75.00$

$\quad N_0 = $ **10075.00000**

E_0 = ?????.?????

$E_0 = TS\ E$

$E_0 = \mathbf{20000.00000}$

—Point 4—-

$Lsb = POS(sta) - TS(Sta)$

$Lsb = 11+50.00 - 10+00.00$

$Lsb = \mathbf{150.00}$

$Cb = Lsb - (0.00034 * a^2 * (Lsb / 100)^5)$

$Cb = 150.00 - (0.00034 * 0.66667^2 * (150.00 / 100)^5)$

$Cb = \mathbf{149.99885}$

$DEFb = (a * Lsb^2) / 60000$

$DEFb = (0.66667 * 150.00^2) / 60000$

$DEFb = \mathbf{0.25000° \text{ or } 0°15'00''}$

$Xb = Cb * Cos(DEFb)$

$Xb = 149.99885 * Cos(0°15'00'')$

$Xb = \mathbf{149.99742}$

$Yb = Cb * Sin(DEFb)$

$Yb = 149.99885 * Sin(0°15'00'')$

$Yb = \mathbf{0.65449}$

$N_4 = ?????.?????$

$\quad Y_3 = Yb + (Cos(DEFb * 3) * W_2) - W_2$

$\quad Y_3 = 0.65449 + (Cos(0°15'00'' * 3) * 75.00) - 75.00$

$\quad Y_3 = \mathbf{0.64806}$

$\quad N_4 = N_0 + Y_3$

$N_4 = 10075.00000 + 0.64806$

$N_4 = \mathbf{10075.64806}$

$E_4 = ?????.?????$

$X_3 = X_b - (Sin(DEF_b * 3) * W_2)$

$X_3 = 149.99742 - (Sin(0°15'00" * 3) * 75.00)$

$X_3 = \mathbf{149.01570}$

$E_4 = E_0 - X_3$

$E_4 = 20000.00000 - 149.01570$

$E_4 = \mathbf{19850.98430}$

—Point 5—-

$L_{sb} = POS(sta) - TS(Sta)$

$L_{sb} = 12+75.00 - 10+00.00$

$L_{sb} = \mathbf{275.00}$

$C_b = L_{sb} - (0.00034 * a^2 * (L_{sb} / 100)^5)$

$C_b = 275.00 - (0.00034 * 0.66667^2 * (275.00 / 100)^5)$

$C_b = \mathbf{274.97623}$

$DEF_b = (a * L_{sb}^2) / 60000$

$DEF_b = (0.66667 * 275.00^2) / 60000$

$DEF_b = \mathbf{0..84028° \text{ or } 0°50'25"}$

$X_b = C_b * Cos(DEF_b)$

$X_b = 274.97623 * Cos(0°50'25")$

$X_b = \mathbf{274.94666}$

$Y_b = C_b * Sin(DEF_b)$

$Y_b = 274.97623 * Sin(0°50'25")$

Y_b = **4.03255**

N_5 = ?????.?????

  Y_4 = Y_b + (Cos(DEFb * 3) * W2) - W2

  Y_4 = 4.03255+ (Cos(0°50'25" * 3) * 75.00) - 75.00

  Y_4 = **3.95997**

  N_5 = No + Y_4

  N_5 = 10075.00000 + 3.95997

  N_5 = **10078.95997**

E_5 = ?????.?????

  X_4 = X_b - (Sin(DEFb * 3) * W2)

  X_4 = 274.94666 - (Sin(0°50'25" * 3) * 75.00)

  X_4 = **271.64796**

  E_5 = Eo - X_4

  E_5 = 20000.00000 - 271.64796

  E_5 = **19728.35204**

—Point 6—-

Lsb = POS(sta) - TS(Sta)

Lsb = 13+00.00 - 10+00.00

Lsb = **300.00**

Cb = Lsb - (0.00034 * a^2 * (Lsb / 100)^5)

Cb = 300.00 - (0.00034 * 0.66667^2 * (300.00 / 100)^5)

Cb = **299.96328**

DEFb = (a *Lsb^2) / 60000

$DEFb = (0.66667 * 300.00^2) / 60000$

$DEFb =$ **1.00000° or 1°00'00"**

$Xb = Cb * Cos(DEFb)$

$Xb = 299.96328 * Cos(1°00'00")$

$Xb =$ **299.91759**

$Yb = Cb * Sin(DEFb)$

$Yb = 299.96328 * Sin(1°00'00")$

$Yb =$ **5.23508**

$N_6 = ?????.?????$

    $Y_5 = Yb + (Cos(DEFb * 3) * W_2) - W_2$

    $Y_5 = 5.23508 + (Cos(1°00'00" * 3) * 75.00) - 75.00$

    $Y_5 =$ **5.13230**

    $N_6 = N_0 + Y_5$

    $N_6 = 10075.00000 + 5.13230$

    $N_6 =$ **10080.13230**

$E_6 = ?????.?????$

    $X_5 = Xb - (Sin(DEFb * 3) * W_2)$

    $X_5 = 299.91759 - (Sin(1°00'00" * 3) * 75.00)$

    $X_5 =$ **295.99239**

    $E_6 = E_0 - X_5$

    $E_6 = 20000.00000 - 295.99239$

    $E_6 =$ **19704.00761**

NOTES

CONCLUSION

Utilizing a step by step approach as outlined in this book, it is fairly easy to solve for a true parallel offset to a centerline spiral curve.

Computer programs have been designed and written that makes easy work for these type of calculations. In order to appreciate the mathematical beauty of spiral curve offset, you are encouraged to manually calculate one every now and then.

Here is a link to exhibits and a Spiral Curve Offset Program.

http://www.cc4w.net/products.html

ABOUT THE AUTHOR
Jim Crume P.L.S., M.S., CFedS

My land surveying career began several decades ago while attending Albuquerque Technical Vocational Institute in New Mexico and has traversed many states such as Alaska, Arizona, Utah and Wyoming. I am a Professional Land Surveyor in Arizona, Utah and Wyoming. I am an appointed United States Mineral Surveyor and a Bureau of Land Management (BLM) Certified Federal Surveyor. I have many years of computer programming experience related to surveying.

This book is dedicated to the many individuals that have helped shape my career. Especially my wife Cindy. She has been my biggest supporter. She has been my instrument person, accountant, advisor and my best friend. Without her, I would not be the professional I am today. Cindy, thank you very much.

Other titles by this author:

http://www.cc4w.net/ebooks.html

www.ingramcontent.com/pod-product-compliance
Lightning Source LLC
Chambersburg PA
CBHW040854180526
45159CB00001B/420